Planets

Dr Emily Drabek-Maunder

Royal Observatory Greenwich
Illuminates

First published in 2021 by Royal Museums Greenwich
Park Row, Greenwich, London SE10 9NF

ISBN: 978-1-906367-82-4

At the heart of the UNESCO World Heritage Site of
Maritime Greenwich are the four world-class attractions
of Royal Museums Greenwich – the National Maritime
Museum, the Royal Observatory, the Queen's House and
Cutty Sark.

rmg.co.uk

A CIP catalogue record for this book is available from
the British Library.

Typesetting by ePub KNOWHOW
Cover design by Ocky Murray
Diagrams by Dave Saunders
Printed and bound by
CPI Group (UK) Ltd, Croydon, CR0 4YY

MIX
Paper from
responsible sources
FSC® C020471

About the Author

Dr Emily Drabek-Maunder is an astronomer, astrophysicist and science communicator. She is currently the Senior Manager of Public Astronomy at the Royal Observatory Greenwich. As an astrophysicist, she used microwave and radio telescopes to study the formation of stars and planets in our galaxy at Cardiff University, Imperial College London and University of Exeter. Emily became interested in astronomy as a child, when she would look up at the night sky and wonder if we were alone in the Universe. She still does this today.

Entrance of the Royal Observatory, Greenwich, about 1860.

About the Royal Observatory Greenwich

The historic Royal Observatory has stood atop Greenwich Hill since 1675, and documents over 800 years of astronomical observation and timekeeping. It is truly the home of space and time, with the world-famous Greenwich Meridian Line, awe-inspiring astronomy and the Peter Harrison Planetarium. The Royal Observatory is the perfect place to explore the Universe with the help of our very own team of astronomers. Find out more about the site, book a planetarium show, or join one of our workshops or courses online at rmg.co.uk.

Contents

Introduction   1

In the Beginning: The Birth of a Solar
System   8

Navigating Our Solar System   23

Small but Mighty: The Rocky Planets   34

Serene Giants: The Outer Planets   54

Everything Left Over: Dwarf Planets,
Comets, Asteroids and Meteoroids   72

Alien Worlds: Our Search for
Exoplanets   84

A Final Note: Is There Life in the
Universe Other Than on Earth?   99

Glossary   103

Introduction

What do you picture in your mind when you hear the word 'planet'? Doubtless, there are as many ideas about what a planet is like as there are planets in our universe. To you, a planet may mean our home, the Earth; that giant blue and white marble. You may imagine the Solar System and the eight planets in orbit around the Sun. Other connotations of a planet may be entirely unknown and mysterious, outside of your everyday experiences and completely alien. The word may conjure up science-fiction fantasies of travelling to different worlds and finding extra-terrestrial life. Perhaps you feel a longing for space exploration and realise how little of our universe, humankind's final frontier, we've been able to explore.

Awareness of the existence of the planets in our solar system dates back several millennia, but planets were originally thought to be very different than we now understand them to be. Since ancient times, five planets have been easily visible to the naked eye – Mercury, Venus, Mars, Jupiter and Saturn. They appear like stars in the night sky, as bright points of light. However, early astronomers noticed that these so-called stars roamed across our skies in a way that other stars did not. The Ancient Greeks called these moving lights *planētai* or 'wanderers', for this reason.

We now know that in our solar system planets move more quickly and more irregularly in the sky relative to distant stars because they are orbiting the Sun, just like the Earth. The heliocentric view of our solar system, with the Sun in the centre and the planets in orbit around it, was not obvious to early astronomers. The spin of the Earth makes it seem like the Sun, planets and stars are rising and setting every day. Many civilisations assumed that this meant the Earth was the centre of the Universe and everything else orbited around it.

Although the ancient Greek philosopher Aristarchus (310–230 BCE) first proposed that the Sun was in the centre of the Solar System, it took around two millennia for this to be the accepted view of our tiny part of space. It wasn't until the invention of the telescope in the 1600s that astronomers could finally start gathering evidence for what our solar system was really like.

Known as the father of observational astronomy, Galileo Galilei (1564–1642) was one of the first people to build a telescope that was fit to observe the night sky. He began studying things in our own solar system – the Sun, Moon and planets – and in doing so, he found the first clues to understanding our corner of the Universe. Galileo's observations of Jupiter were astonishing at the time. With a telescope weaker than most modern binoculars, he managed to not only observe Jupiter, but four moons in orbit around the planet. This was a groundbreaking discovery. If moons orbited Jupiter, then this meant that not everything in space was orbiting the Earth. Galileo gained further evidence of this hypothesis by observing the planet Venus, discovering that it had similar phases to Earth's moon. The only explanation for Venus having phases in this manner was if the planet were in orbit around the Sun, and closer to the Sun than the Earth.

While Galileo's work changed our understanding of the cosmos, it also had serious consequences for him personally. In demonstrating that the Sun was at the centre of our solar system, Galileo was tried for heresy during the Roman Catholic Inquisition and was found guilty. He was not only put under house arrest for his observations of the Universe, but he was forced to renounce his work. It took another two centuries for the heliocentric model of our solar system to be accepted as fact and even longer for Galileo to be acquitted of his 'crimes'.

Over the past millennia, since humans started looking up at the stars and charting their locations, our knowledge and understanding has progressed incredibly. Not only have we sent humans to walk on the Moon, but we have sent spacecraft to study every planet in our solar system up close. We have been able to use telescopes to peer into our starry sky and find other solar systems with their own planets. We've seen planets as they first begin to form and analysed their atmospheres to ascertain the possibility of these distant worlds sustaining life.

While we know a lot of information about planets and our solar system, *we do not know everything*. Astronomy is a humbling branch of science; it highlights how small and fragile we are and how little we know about our vast universe. However, we live in an exciting and ambitious time for science, where discoveries are made every day and humans push the boundaries of knowledge ever further. Although we don't know everything, what we have been able to understand and uncover is truly remarkable for beings from a little planet out there in the great cosmos.

In this book, you'll learn what it takes for a planet to form in our galaxy. There will be a tour of the Solar System, including work done by past astronomers who were the first to study the planets with telescopes. You'll visit each planet in our solar system and discover unique features on each of these worlds. The dwarf planets and leftovers from the formation of our solar system are described towards the end of the book, as well as an explanation of what makes a planet different to a dwarf planet. Lastly, we'll leave our solar system behind to explore exoplanets and other solar systems in our galaxy.

In the Beginning:
The Birth of a Solar System

When we look up at the sky at night, all of the stars we see overhead are in our galaxy, the Milky Way. There are trillions of galaxies out there in space, but the Milky Way galaxy is our home. It contains hundreds of billions of stars and may contain as many planets. However, our solar system only contains one star, the Sun.

All stars start out in regions inside our galaxy called nebulae, which comes from the Latin *nubes*, or 'clouds'. While clouds in space may sound unusual, these nebulae do actually look a little like bright, rainbow-coloured clouds and they are some of the most beautiful and extraordinary places that can be found in space. Take a look at the image section for telescope observations of nebulae for examples of these breathtaking regions of our galaxy. Well-documented and stunning nebulae include the Horsehead Nebula, the Orion Nebula and the Carina Nebula.

Nebulae are a bit like stellar nurseries, in which hundreds or even thousands of stars are being nurtured. The stars that we see in our night sky formed millions to billions of years ago, and new stars are forming inside our galaxy all the time. By studying these young stars in their parental nebulae, we can better understand how all stars formed, including our own Sun. We can also get a better idea for how solar systems with planets form around these stars.

In general, nebulae are cold, cloud- or mist-like regions in space made up of dust and gas. The gas found in nebulae is mainly molecular, or gas that is made up of two or more atoms bonded together. While the most common molecular gas in our universe is molecular hydrogen (H_2), larger and more interesting molecules can grow inside a dense nebula. Inside nebulae are molecules that we need to live on Earth, like water vapour (H_2O). Other molecules are poisonous to us, like carbon monoxide (CO) and hydrogen cyanide (HCN) gases. There is also an abundant supply of alcohol in space, in the form of ethanol (C_2H_6O), in far larger amounts than we can brew it on Earth! Some of the more unusual and complex molecules that can be found include ethyl formate ($C_3H_6O_2$), which has the unique property of tasting like raspberries and smelling like rum[1]!

---

[1] If anyone ever asks you what space tastes and smells like, then you can officially say 'rum and raspberries.' I just wouldn't recommend *trying* to taste and smell space due to all of those other poisonous gases mentioned previously.

While these nebulae might look like tranquil regions in space, the reality is that forming a star is a violent and turbulent process. Dust and gas inside nebulae move at tens to thousands of kilometres per second. The reason nebulae appear relatively peaceful in the images we have is because they are viewed at a vast distance from our solar system. One of the closest nebulae to us – the Rho Ophiuchi cloud complex – is around 400 lightyears in distance away from the Earth. This means that the light we see from the stars in Rho Ophiuchi, where light travels at around 300 million metres per second, takes a little over 400 years to reach our telescopes and cameras here on the Earth. It takes thousands of years for starlight to reach us from some of the furthest nebulae in the Milky Way galaxy.

Speaking of distances, the large distance scales found in space means we are literally looking back in time when we look out at the Universe. The light from stars and other galaxies can take hundreds, thousands, millions and even billions of years to reach us, which means that we are only seeing what these objects looked like in the past. Who said time travel isn't real?

To form a planet, first you need a star, and for a star to form, you need gravity to act on the 'stuff' inside the nebula. The material inside a nebula is not uniformly distributed, it's relatively clumpy, so gravity will begin to pull some of the dust and gas together in these clumpier areas of a nebula. The collapse of this material under gravity is what forms a baby star, or what astronomers call a protostar.

A protostar is *not* the same thing as a star, but it is on its way to becoming one. One of the main differences between a protostar and a star is temperature. The **natal nebula** of a star is extremely cold, around -266 to -243°C, which is only 7 to 30 degrees above **absolute zero**. As a clump collapses to form a protostar, gravity is converted to kinetic energy which causes the dust and gas to heat up by 2,000 to 3,000 degrees! Over time, the protostar heats to around 15 million degrees, the temperature needed for nuclear fusion[2], and finally becomes a star. From absolute zero to millions of degrees, the journey a star makes from birth is extreme.

The protostar itself is stable and in balance, meaning it is held together by gravity and supported internally by pressure. One unique feature of protostars is that they are still embedded

---

[2] Nuclear fusion is what occurs in all stars, which are known as 'main-sequence stars' in astronomy. This is a process where atoms are smashed together and converted into other atoms at high temperatures and pressures, deep in the bellies of stars. When stars start out, hydrogen is fused together to form helium. Fusion is quite an unusual concept. Think of it as the equivalent to smashing a few cars together and getting a bigger, brand new car afterwards!

in their parental nebula, which makes them difficult to observe. In optical light (light we can see with our eyes), the shroud of dust and gas surrounding the protostar completely obscures our ability to see them. For astronomers to study the very early stages of star and planet formation, less energetic light from outside the visible spectrum must be observed from these nebulae, mainly infrared, microwave and radio light.

Planets can only begin to form once a protostar exists. Over time, surrounding material will continue to fall towards the spinning protostar, helping it to gain more mass[3]. As the surrounding dust and gas falls toward the protostar, it is swept up in orbit around the star and flattens out into a disc-shape, due to **angular momentum**.

---

[3] Not all the material will become a part of the protostar. Some of the dust and gas will be pulled into massive, powerful outflows that flow from the poles of the star.

The result is the protostar in the centre, surrounded by a disc made of dust and gas.

This might seem like a surprising phenomenon, but protostars with their surrounding dust and gas are a little like pizzas of space. Imagine making a pizza base. The base starts out a spherical lump of dough. To make a pizza, the ball of dough is thrown up into the air with a little bit of a twist. Once you do this a few times, the ball of dough slowly flattens out into a disc. This process is also caused by angular momentum. In the case of star formation, the disc-shape emerges when surrounding material is pulled back to the central protostar by gravity, but that pull is resisted by its movement around the protostar. Just like a ball of dough flattens into a pizza, the material orbiting protostars will flatten out into a disc.

The discs in orbit around protostars are gold dust to astronomers because they are where the building blocks of planets can be found. If humans ever hope to uncover how solar systems, planets and life form in our universe, then these discs may hold the key.

Everything in a solar system forms from these discs, which orbit young stars, including planets, dwarf planets, comets, asteroids and moons. To understand how planets form, we need to first understand some of the smallest things in space. Nebulae are made up partly of dust, which it is nice to think of as star dust. Before you start to imagine a ball of fluff floating around in space, similar to what you might vacuum in your home, think again. Star dust is more like smoke, such as cigarette smoke or the smoke from a fire. This type of dust is made up of small grains of silicon, carbon or iron which combine with oxygen to form minerals. These cosmic dust grains are incredibly tiny, as small as a micrometre or even a nanometre in size (one-millionth to one-billionth of a metre!). To put that into perspective, this means a dust grain in space is 10 to 10,000 times smaller than the width of a human hair.

So, why should we care about something that is so small? Well, over 2,000 years ago, an ancient Roman poet called Horace, described the transience of our lives and the finality of death with the phrase *'pulvis et umbra sumus'*. If you've not brushed up on your Latin recently, this translates to 'we are dust and shadow.' Astronomically speaking, he wasn't wrong. Many things around us, including the cosmic bodies we can observe in space, are actually made up of tiny dust grains. Dust grains made it possible for life, as we know it today, to exist on Earth.

Over millions of years, dust grains in the disc around a newly formed star will begin to stick together and form into small, centimetre-sized rocks or 'pebbles'. These pebbles stick together to form bigger rocks and these keep growing until we have something that stretches to several kilometres in size, known as a **planetesimal**. These larger objects will be able to use gravity to pull even more material towards them until they have grown into something the size of a planet.

The process of planet formation is particularly dramatic due to the amount of debris orbiting a star during this period. At high speeds, collisions can completely destroy growing planets. If the planet survives this chaotic time, the pressure in the planet's core will increase and the temperature will rise, which causes the materials to melt together and separate into different layers.

Planets obtain their atmospheres in different ways. More massive planets, like Jupiter or Saturn, will get most of their mass and atmospheres from gas being attracted to the planets by gravity. Smaller planets, like Venus, Earth or Mars, may get their atmospheres from outgassing (for example, through volcanic eruptions) or from impacts by comets or asteroids. This is just one example of the differences between the inner and outer planets in our solar system.

From gaining an understanding of planet formation, astronomers now know that all of humankind is literally living on a giant ball of star dust that stuck together over millions of years. Truth is most certainly stranger than fiction!

Although forming a planet sounds like a simple process, researchers are *still* trying to work out the intricate details of the process, including how the planets in our own solar system formed. The mechanism for sticking the dust grains together and growing them into larger pebbles and rocks is difficult to study due to the sheer size of our galaxy and the long timescales of planet formation. However, astronomers can explore these current unknowns in a few different ways.

Firstly, the conditions of space can be recreated in a laboratory to observe how dust grains might stick together in a disc orbiting a star. While larger grains of dust will be attracted by gravity, it is the behaviour of the smaller dust grains, too light to be affected much by gravity, that is still puzzling. Current research suggests that these grains may develop electrical charges and stick together to form larger grains. This is similar to static electricity on the Earth, which is what happens when you rub a balloon in your hair and your hair is attracted to the balloon.

Another method for investigating how dust sticks together over time is to study rocks that come from space. Meteorites are made up of solid debris from space that were originally pieces of asteroids, comets or meteoroids that fall through our atmosphere and hit the Earth. The majority of these meteorites are **chondrites**, meaning they are made up of chondrules, or small, round particles. These chondrite meteorites give an insight into how larger bodies grow from smaller grains in space, eventually leading to planets.

Lastly, powerful telescopes can be used to observe discs. While the mechanism for forming pebbles in space is still not well understood, telescopes that are able to see light at radio wavelengths can observe where these pebbles are forming in orbit around protostars to give a better idea of how they form. Telescopes studying planet formation include the e-MERLIN telescope, which is an array of telescopes found across England[4]. In the future, the Square Kilometre Array (SKA), which will be the world's largest radio telescope once

[4] While you might think that the stereotypically rainy weather in the United Kingdom isn't very well suited for studying the stars, radio telescopes can be used in poorer weather conditions because radio light can pass directly through the clouds. Perfect for soggy climates!

construction is complete, will also have the capability to study planet formation in previously unseen detail and spot baby planets as they grow.

Navigating Our Solar System

Even though scientists have an idea about the process of star and planet formation in other solar systems throughout our galaxy, how can we use this information to understand our own solar system? One clue is in the structure of our solar system, which is a direct result of its formation and gives us some clues into how our solar system formed and evolved.

Our solar system consists of a star (see Figure 1), which is our Sun, and eight planets[5] as well as moons, comets, asteroids and dwarf planets. The four inner planets – Mercury, Venus, Earth and Mars – are called the terrestrial or rocky planets. These planets are mainly made up of rocks and metals, and their defining feature is that they have surfaces that we can stand on, theoretically at least. In reality, you wouldn't necessarily *want* to stand on many of these planets because of the harsh conditions that exist on their surfaces (more about this in the next section). The four outer giant planets are gas giants Jupiter and Saturn, and ice giants Uranus and Neptune. Gas and ice giant planets do not have easily defined surfaces and the gas extends to the inner portions of these planets.

---

[5] Even I struggle not to write that there are nine planets in the Solar System. Pluto is not left out of this book. Dwarf planets are important to understand in context with the rest of the Solar System.

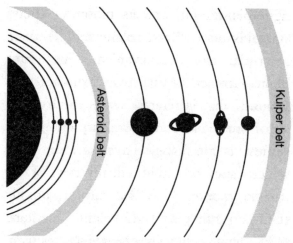

Figure 1: The layout of the Solar System, beginning with the Sun. Following the Sun are rocky planets Mercury, Venus, Earth and Mars. After Mars is the asteroid belt and then the giant planets Jupiter, Saturn, Uranus and Neptune. The Kuiper belt is beyond the orbit of Neptune, which is where Pluto can be found.

The rocky planets are found closer to the Sun, the gas giant planets are further out, and the ice giant planets are the furthest away. It is no coincidence that the planets are grouped in this manner. From studying much younger solar systems in their parent nebulae, astronomers can see that rocky materials and metals are the main survivors of the formation process closer to young stars because of the higher temperatures at these smaller distances. The giant planets formed further away from the Sun, beyond what is referred to as the **frost or snow line**. This is an imaginary line which represents the distance from the Sun after which gases or liquids, like water (H_2O), carbon monoxide (CO), carbon dioxide (CO_2), ammonia (NH_3) and methane (CH_4), will begin to freeze into ice grains.

The frost line in our solar system can be found between the orbits of Mars and Jupiter. The planets beyond the frost line likely started out with cores made of both rock and ice and were able to accumulate massive atmospheres made from hydrogen and helium gases, which are lighter and more abundant elements. Stars are similarly made from hydrogen and helium because they are the most abundant elements in the Universe. Interestingly, the composition of the Universe wasn't known until relatively recently. In 1925, astrophysicist Cecilia Payne-Gaposchkin (1900–1979) discovered that stars are composed of the lightest elements. However, her work was originally doubted because it went against the convention at the time that there was no difference between the make-up of the Earth and the Sun. While her work was subsequently confirmed, she is often not given the appropriate credit for her discovery of a fact so fundamental to our universe.

Not only is there a difference in the types of planets as you move further away from the Sun, but the planets also travel at different speeds. Knowing that the speed of the planet changes with its distance from the Sun meant that astronomers hundreds of years ago were able to calculate the size and shape of our solar system before Cecilia Payne-Gaposchkin ever figured out the materials that made up our Sun.

Astronomer Johannes Kepler (1571–1630) began work on understanding the movement of the planets in the early 1600s, particularly how the Earth and the other five planets known at the time – Mercury, Venus, Mars, Jupiter and Saturn – orbited the Sun. In this time-period it was still widely thought that the Earth was at the centre of the Solar System and the Sun and other planets orbited the Earth. Alongside Galileo's observations of the planets, Kepler's work was an attempt to directly determine the nature of the Solar System.

Kepler started out with fellow astronomer Tycho Brahe's (1546–1601) observations of the planet Mars, studying how Mars moved across the backdrop of constellations in the sky. From these observations, he discovered that the orbits of the planets around the Sun are not perfect circles, but elliptical or oval in shape (as seen in Figure 2). This means that as they orbit, planets are sometimes closer to the Sun and sometimes they are further away.

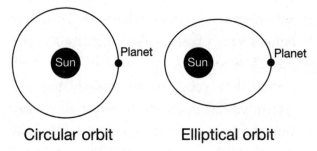

Figure 2: A comparison between a planet with a circular orbit and an elliptical orbit.

After extending his study to observations of the other planets, Kepler made a breakthrough when he realised that planets at distances further away from the Sun orbited more slowly. This discovery even surprised Kepler himself, who wrote in 1619, 'I first believed I was dreaming.' It was as if there was a force attracting the planets to the Sun, which became less powerful with distance. While we now

29

know this unseen force attracting planets to stars is gravity, Kepler's work laid the foundation for Sir Isaac Newton (1643–1727) to write his laws on gravity over a half a century later and for the distance between the planets to be calculated in the early 1700s[6].

It is important to remember that the full three-dimensional shape of our solar system is a disc even though the orbits of the planets around the Sun are elliptical. This means that if you look at the Solar

---

[6] Kepler was not only a brilliant scientist, but he was also one of the world's first science fiction authors. His work *Somnium*, or *The Dream*, features creatures who live on the Moon. With the change in perspective, these creatures think the Earth is orbiting the Moon instead of the Moon orbiting the Earth. Kepler originally designed this work to be an allegory. Even though it looks like the Sun orbits the Earth, the Earth is actually orbiting the Sun. Sadly, this work was mistaken as something more than fiction and it may have contributed to a trial against Kepler's mother for witchcraft.

System from the side, it is flat like a pancake. The shape of the Solar System is connected to the young stars astronomers have been studying, these protostars with discs of dust and gas. This is clear evidence that everything in our solar system orbiting the Sun evolved over millions and billions of years out of the Sun's disc. The icy and rocky debris that can be found in our solar system is made up of the leftover remnants of this disc, composing the asteroid belt in between the orbits of Mars and Jupiter and the Kuiper belt beyond the orbit of Neptune.

With modern telescopes, Kepler's work on the motion of our solar system planets has now been verified for other solar systems in our galaxy. Astronomers have even been able to uncover the detailed structures of discs with planets as they begin to form. Surprisingly, many of these discs look like large rings around their stars, with noticeable gaps in them. These rings are expected to form once planets begin to grow from the dust and gas in the disc. Planets can carve gaps into the disc and create ring-like structures, like the rings of debris found in our asteroid belt and the Kuiper belt.

While humans have studied our solar system for centuries, we have only in the past few decades been able to send spacecraft to study planets directly. Telescopes can offer us a piece of the overall puzzle, but we need direct measurements to understand these environments better. The size of our solar system has meant that some of these uncrewed space missions have lasted multiple generations. To this day, there has never been a spacecraft that has travelled outside our solar system, though there are some spacecraft that are well on their way to achieving this goal. For example, *Voyager 1* and *Voyager 2* were launched into space in 1977 with the goal of studying the outer planets of the Solar System and have been travelling further out into space ever since. Both spacecraft are now far beyond the orbits of the planets and are currently in interstellar space[7].

---

[7] To communicate with *Voyager 1* and *Voyager 2*, it takes around 20 hours for a radio signal to travel from the spacecrafts back to the Earth.

Small but Mighty: The Rocky Planets

While they may be the smallest planets in our solar system, the rocky planets have some of the most diverse environments. From scorched to frozen worlds, the first four planets closest to the Sun all have surfaces that can be explored. Surprisingly, not all of these planets have moons. Mercury and Venus do not have any orbiting companions. While the Earth has one moon, conveniently named *the* Moon, Mars actually has two. The moons in orbit around Mars are named Phobos and Deimos, ancient Greek for 'terror' and 'dread'. Let's find out a bit more about the rocky planets, from the smallest and closest one to the Sun, Mercury, through to Mars, the last rocky planet before the asteroid belt.

Mercury

Following Pluto's change in classification to dwarf planet in 2006, Mercury became the smallest planet in our solar system. At 4,880 kilometres in diameter, this planet is just 1.5 times the size of the Moon and looks pretty similar to it. Unlike the other rocky planets, Mercury has little to no atmosphere because its small size means that it doesn't have enough gravity to keep hold of one.

While Mercury might initially seem like a dull place, don't underestimate this little planet. The thin atmosphere it does have is called an **exosphere,** which is made up of atoms that are blown off the surface of the planet by the Sun's light.

Different gases, like hydrogen, helium, oxygen, sodium, calcium, potassium and even water vapour can be found in this exosphere. Unfortunately, the exosphere isn't stable, and over time the gases are swept away fully by the Sun and replenished from atoms on the surface of the planet once more.

As the closest planet to the Sun, Mercury is continuously blasted by solar radiation. When standing on Mercury's surface, the Sun would look about three times larger in the sky than on the Earth. The Sun's radiation causes the planet to heat up to 430°C in daylight. However, since the atmosphere of the planet is thin, there is no way to hold onto this heat. This means that while the day-side of the planet is hot, the night-side is incredibly cold, with temperatures as low as -170°C.

Its proximity to the Sun also means that Mercury orbits the Sun quickly.

A year on this planet is the equivalent to 88 days on the Earth, so theoretically you would be four times older on Mercury and it wouldn't be uncommon for people to reach their 200th birthday! The planet's swiftness in orbiting the Sun is reflected in its name, as in Greek mythology, Mercury was the Roman messenger god who carried messages between Mount Olympus (the 'heavens') and the Earth. Although a year passes quickly on the planet, the days are much longer. Due to its slower spin but fast orbit, a day on Mercury lasts 176 Earth-days, which is actually longer than Mercury's year[8]!

---

[8] Mercury's *solar day* is equal to 176 days on the Earth. This is not the same as a *sidereal day* which is the amount of time it takes for Mercury to spin once. A solar day is the average length of time it takes to see the Sun at its highest point in the sky from one day to the next. So, a day on Mercury will feel like 176 Earth days in length.

Of all of the rocky planets, we know the least about Mercury, mainly due to its proximity to the Sun. Two satellites have studied the planet up close, NASA's *Mariner 10* and *Messenger* spacecraft, which provided some of the first detailed images of the surface of the planet. With the arrival of new missions, including BepiColombo in 2025, astronomers will soon better understand how planets close to their stars form through investigating Mercury's surface, interior, atmosphere and magnetic field.

Astronomers can also study Mercury when, from our perspective on Earth, it occasionally passes in front of the Sun. This is known as a transit. The only visible transits from Earth are those of the planets Mercury and Venus because they are the only planets that orbit the Sun within Earth's orbit. Mercury transits the Sun about 13 times every century, normally in May or November. Recent transits were in May 2016 and November 2019, but the next transit of Mercury isn't expected until 13 November 2032. When Mercury passes in front of the Sun, astronomers can explore the exosphere that surrounds the planet and relate this knowledge to exoplanets (planets orbiting other stars in our galaxy). Exoplanets are explored more in the later sections, particularly their transits.

Venus

Named after the ancient Roman goddess of love and beauty, the planet Venus has previously been called Earth's twin or sister planet due to its similar size, mass and distance from the Sun. This planet can sometimes be seen in the early morning or evening with just the naked eye, when it is the brightest object in the sky except for the Moon. This is why Venus is also sometimes referred to as the morning star (Phosphoros); or the evening star

(Hesperos). Early astronomers originally thought that Venus was two different planets.

While Venus might seem like Earth's twin, in reality the planet is very different from the Earth and a great deal more hostile to humans. When the first spacecraft were sent to Venus in the 1960s they uncovered a hellish landscape. Although Venus is twice as far away from the Sun as Mercury, the thick yellowish clouds surrounding this planet trap heat near its surface. This makes the average temperature on Venus much hotter than an oven and the hottest planet in our solar system, boasting highs of 460°C.

The question you should be asking is not *if* a person could die on this planet, but *how many ways* could they die. While the heat is an obvious hazard for life on Venus, another threat to human life is the pressure on its surface, which is 92 times that of Earth's. Standing on Venus would feel as if you were 900 metres underwater, experiencing pressure high enough to crush the human body. There is also the small matter of its clouds, which are made from sulfuric acid...

None of these threats compare to the global resurfacing events that may occur on Venus. With the exception of Mars, Venus is the only other planet in our solar system to have probes successfully land on its surface. Throughout the 1970s to 1990s, the data and photographs sent back to the Earth from the Soviet Union's *Venera* probes, alongside NASA's *Magellan* spacecraft, showed Venus as a haunting desert broken up by craters and volcanoes. However, the data sent back from these spacecraft indicated that the surface of the planet seemed relatively young compared to the age of our solar system. This has led astronomers to theorise that the surface of Venus was renewed a few hundred million years ago. Venus likely does not have plate tectonics like Earth, meaning that the surface of the planet is fixed and does not gradually move about. Without plate tectonics, heat builds up under the surface of Venus, weakening its surface or crust. Eventually, lava forces its way up to cover the old crust of the planet and cools to form a new surface.

While the possibility of life on this planet seems nearly impossible, there has been continued speculation about life existing in the atmosphere of Venus. At altitudes 50 to 60 kilometres above the planet's surface, temperatures are a much milder -20 to 65°C. Microorganisms or microbes could potentially thrive in these environments, using sunlight as a fuel source. Only time will tell if life exists on this planet, when future spacecraft, designed to search its upper atmosphere for evidence of creatures that can live in these harsh conditions, eventually make their way to Venus.

Earth

The largest of the rocky planets is the only known planet with a diverse variety of life. From single-celled organisms to complex organisms such as humans, life is found everywhere on our planet. But what is it that makes Earth so different compared to the other rocky planets? Why is our planet so special?

There are a variety of reasons why Earth is favourable to life and they all mainly depend on the presence of liquid water. Earth is unique compared to the other planets in our solar system because of the abundance of liquid water that is found on its surface. Every form of life on Earth needs water to survive, since it is necessary for the chemical reactions occurring inside cells.

This connection between life and liquid water has guided astronomers when defining the habitable zone around a star. The habitable zone, or the Goldilocks

Zone, is the area around a star where the temperature is mild enough for water to exist as a liquid on the surface of an Earth-like planet. In other words, the temperature isn't too hot or too cold – it's 'just right!' In our solar system, the boundaries of the habitable zone are difficult to identify, but generally it is thought to begin around the orbit of Venus, encompassing the Earth, and end around the orbit of Mars. In practice this does not mean that life exists everywhere within this region, or even that liquid water is present, but only that the temperature is more favourable for water to exist as a liquid. Other things will affect whether water is present on a planet, like the pressure in a planet's atmosphere or heating from greenhouse gases (which, incidentally, is why Venus is as hot as it is).

The astonishing thing about the Earth is that not only is life found, but it thrives *everywhere*. Life can be found even in the harshest conditions, from boiling temperatures at the bottom of the ocean near volcanic vents to the acidic liquids that drain from metal or coal mines. These forms of life are known as **extremophiles,** named after the extreme environments in which they live, where it was previously thought to be impossible for life to survive.

Some extremophiles on Earth are nearly impossible to kill. For example, *Tersicoccus phoenicis* is a bacterium that has only ever been found in two places on the Earth, both times in clean rooms designed to build spacecraft! These clean rooms were designed to have minimal amounts of outside contaminants, such as dust and microbes. The air to the rooms is filtered, surfaces are cleaned with alcohol and hydrogen peroxide, items are heated to high temperatures, and all the staff do their work in protective suits. However, *Tersicoccus phoenicis* still managed to survive in these environments. These extremophiles are so indestructible that astronomers think it is possible that some of them may have accidentally hitched a ride to Mars on past missions, like NASA's *Curiosity* rover. It is therefore possible that we may have already accidentally seeded life on other planets.

Understanding these extremophiles on the Earth will give us a better idea of the kind of life that may exist in other places. While Earth is the only planet in our solar system that is firmly centred in the habitable zone, extremophiles give us hope that life could exist on Venus or Mars, which are situated on the boundaries of the zone, or in more unexpected environments in our solar system[9].

---

[9] Planets aren't the only environments that may harbour life. Moons in our solar system, particularly some of the moons orbiting Jupiter and Saturn, are known to have liquid water oceans. There will be more on this later.

Mars

Otherwise known as the Red Planet, Mars is named after the ancient Roman god of war. Don't let the violent name fool you, this planet is inhabited entirely by peaceful robots. You read that right – the only 'life' currently known on Mars consists of the landers and rovers that we've been sending there since the 1970s.

The fiery red colour of Mars is not as a result of its temperature, but is caused by the rust (iron oxide) that covers the surface of it. Since Mars is further away from the Sun than the Earth, the planet is significantly colder than our own. An average day on Mars is -60°C. However, the temperature depends on the location on this planet and the season. Although Venus may be known as Earth's sister planet, Mars has seasons which are more similar to those on Earth, caused by the tilt of the planet. During winter the temperature at the poles of Mars can drop as low as -140°C, but in summer on the equator it can be a balmy 35°C!

The landscapes that can be seen on Mars are sprawling deserts, interspersed with canyons, extinct volcanoes and craters. The poles of the planet are made up of carbon dioxide and water ice, and their whiteish colour can sometimes be distinguished even through the telescopes of amateur astronomers. The highest volcano in our solar system is found on Mars, called Olympus Mons, which is 2.5 times higher than Mount Everest and wide enough that it would engulf most of Great Britain.

Mars is the most explored planet out of all the planets in our solar system, owing to its close proximity to Earth and (relatively) mild climate. It is also the only planet that has been explored by rovers which can be driven around remotely by scientists, safe on the surface of the Earth. From these missions, scientists have found increasing evidence that Mars may have had conditions more favourable to life in the past. Liquid water was likely to have existed on the surface of Mars millions to billions of years ago. Certain rocks and gemstones, like opal, have been found on the surface of Mars, that often require liquid water to form. Similarly, features on its surface suggest water may have once flowed on Mars, including large channels emerging from some of the planet's canyons.

Surprisingly, these dry channels on Mars caused significant controversy when they were first observed by Giovanni Schiaparelli (1835–1910) in the 1800s. Astronomers are human, after all, and are not exempt from having runaway imaginations. Schiaparelli described the series of features across the surface of the planet with the Italian word *canali*, which translates to 'channel' or 'canal'. The differences between the translation are subtle, but channels are naturally occurring features on land and canals are engineered. It is easy to see how once an idea is suggested, it can take root and turn into something very different than originally intended. During the turn of the 20th century, there became increasing speculation that intelligent life existed on Mars and the 'canals' were a way for that life to transport water across its surface! Around this time, H.G. Wells published the book *The War of the Worlds*, drawing off these ideas of life on Mars. This was just one of a flurry of science-fiction works involving life on Mars and the existence of Martians. The idea of canals cutting across the surface of the planet was eventually shown to be a clever optical illusion by astronomers Annie Maunder (1868–1947) and Edward Walter Maunder (1851–1928), at the Royal Observatory in Greenwich, London. They enlisted students from the

Royal Hospital School in Greenwich to do experiments that could demonstrate this illusion. These experiments showed that irregular features viewed from a distance appear as though they are straight lines, or canals designed by some intelligent being, but with more powerful telescopes, these canals have been revealed as natural features of the surface of the planet.

Standing bodies of water were likely present on Mars in the distant past, and there is still a possibility liquid water exists on the planet today. The use of radar has shown evidence of lakes below the surface of the planet near the poles. It is thought that liquid water could potentially exist further underground where it is warmer and where the low pressure in the atmosphere won't cause it to evaporate. Missions launched in the 2020s, including those from NASA and the European Space Agency (ESA), will further explore the potential for life on Mars, in both the past and the present.

Even though Mars is one of the closest planets to the Earth, these missions will still between six to eight months to reach the planet[10]. Preparations are being made to send astronauts to Mars, but there are many uncertainties that need to be resolved before this happens. At this point in time, there isn't an existing rocket that can send a spacecraft large enough to carry humans and their supplies all the way to Mars. There is also the matter of preparing astronauts for the Martian environment and years of isolation. Lastly, the technology to get the astronauts back to the Earth afterwards also needs to be developed. A mission of this kind will take decades of preparation and many years to complete. So far, NASA has started

---

[10] For reference, it only takes three days to get to the Moon!

making the first steps in developing new technology needed for a crewed mission to Mars by developing the Space Launch System, a rocket powerful enough to get us there.

Serene Giants:
The Outer Planets

Contrary to popular belief, you will not sink through a gas giant or ice giant planet if you attempted to stand on it. Even though these planets do not have defined surfaces, they are expected to have some sort of metallic or rocky core. The giants of our solar system have the most moons out of all of the other planets, totalling over 150! This section will begin with Jupiter and go in distance order to the last planet, Neptune. Some of the more notable alien moons are also mentioned in more detail.

Jupiter

Jupiter is named after the king of the gods in Roman mythology, an appropriate name for the largest planet in our solar system. This planet is so large that over 1,300 Earths could fit inside it. Actually, *all* of the other seven planets could fit comfortably inside Jupiter!

As a gas giant planet, Jupiter is mainly made up of hydrogen and helium gases. Even if there was a surface to stand on, it wouldn't be recommendable. The temperature inside this planet is estimated to be 35,700°C! The atmosphere of Jupiter has noticeable clouds that surround the planet in bands of brown, white, red and orange. These clouds are made of different elements, including water droplets and ice and ammonia crystals that reflect sunlight in different shades. Throughout these bands are vortices; circular structures in the atmosphere that rotate, caused by streams of gas. A number of storms exist on Jupiter, some of which are temporary, lasting only three to four days on average. These storms are similar to thunderstorms on Earth, but with lightning and large, clump-like clouds that can be thousands of kilometres in size, comparable to entire countries.

Of all of the features on Jupiter, the most well-known is the Great Red Spot, a large reddish vortex that is just south of the equator. This storm has lasted for over 350 years and is currently a third larger than Earth. Although the storm seems stable, the Red Spot has actually decreased in size by around 930 km per year. Astronomers aren't quite sure if the storm will disappear entirely, although if it keeps shrinking at its current rate it could be gone within 20 years.

What lies below the planet's atmosphere is a bit of a mystery, one that has been a focus of study by NASA's *Juno*

'Earthrise' – the rising Earth about five degrees above the lunar horizon taken from the Apollo 8 spacecraft, 24 December 1968.
NASA

The stellar nursery called the Carina Nebula, specifically the Mystic Mountain region, taken by the Hubble Space Telescope.
JPL/NASA/ESA/STScI

Baby stars forming 1,500 lightyears away in the Orion
Nebula, taken by NASA Spitzer Space Telescope and the
Hubble Space Telecope.
JPL/ NASA/JPL-Caltech/STScI

The Horsehead Nebula in the Orion constellation.
JPL/ NASA/ESA/STScI

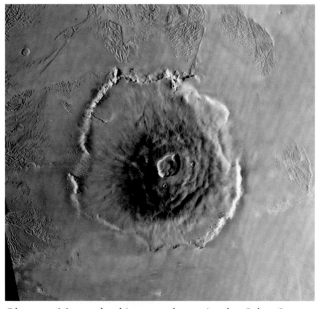

Olympus Mons, the biggest volcano in the Solar System, taken from Viking 1 while orbiting the planet Mars on 22 June 1978.
NASA/JPL

An image of Jupiter showing the Great Red Spot, taken by the NASA/ESA Hubble Space Telescope on 25 August 2020.
NASA, ESA, A. Simon (Goddard Space Flight Center), and M. H. Wong (University of California, Berkeley) and the OPAL team

A monochrome image of Saturn's north polar hexagon and vortex along with its expansive rings.
NASA/JPL-Caltech/Space Science Institute

An enhanced colour image detecting differences in the composition and texture of Pluto's surface, showing what looks like a heart on the surface of the dwarf planet. *NASA/JHUAPL/SwRI*

spacecraft[11]. This craft was designed to better understand how Jupiter was formed by observing the atmosphere and interior of the planet, as well as studying its magnetic fields and its **aurorae**; northern and southern lights similar to the aurorae found on the Earth. *Juno* has uncovered that Jupiter does not have a solid rocky core, but the core is a sort of fuzzy ball without a clear boundary between itself and the other layers of the planet. This was an unexpected find, but one that will help determine how Jupiter formed out of the dust and gas from which it was created.

The larger the planet, the stronger its gravitational pull and the more easily it attracts objects into its orbit. While the Earth may only have 1 moon, the planet Jupiter has 79! Many of these moons are

---

[11] The *Juno* spacecraft is named after Jupiter's jilted wife Juno in Roman mythology. The craft was designed to be a sort of 'spy' on the planet to uncover its 'true nature'.

their own little worlds and are fairly similar to planets. For example, Ganymede and Europa are covered in layers of ice and are thought to hold liquid water oceans below their surfaces. Even though these moons experience freezing temperatures, they are likely heated by tidal friction. The moons are stretched and squeezed by Jupiter's gravity, which heats them up and causes ice below their surfaces to melt. In contrast, the moon Io is the most volcanically active object in our solar system, possessing over 400 volcanoes. The volcanoes make the moon look a little like mouldy cheese, its base yellowish colour coming from the sulphur that makes up its surface. Although Io might look like cheese, its sulphuric smell would actually be more like rotten eggs[12].

---

[12] Remember how Galileo was tried as a heretic for his drawings of Jupiter and its moons? The four moons that Galileo saw through his telescope are named after him, called the Galilean moons. These moons are Io, Europa, Ganymede and Callisto.

Saturn

Saturn has an iconic appearance thanks to its spectacular rings. It is considered by many to be the jewel of our solar system. Like Jupiter, Saturn is a massive gas giant and the second largest planet in the Solar System. It was also named after a Roman god, after the father of the god Jupiter.

Even though all the giant planets in our solar system have faint rings, Saturn's rings are by far the most striking. These rings may look solid, but they are actually made of millions of pieces of rock (carbon) and ice, from the size of grains to as large as houses. The rings extend 282,000 kilometres from the planet, but they are only tens of metres in thickness. This makes the rings appear razor-thin from a distance. The origin of these rings is still one of the many unknowns about our solar system, though astronomers think they could have formed from the pieces of a destroyed moon or from the leftovers of the disc that formed our solar system.

Saturn is mainly composed of hydrogen and helium, but its pale yellowish hue comes from the ammonia crystals in its thick atmosphere. Like the other giant planets, Saturn has no definite surface, though it may have a rocky core. One surprising thing about this planet is that its density is less than that of water, which means that it could technically float if a large enough body of water could be found to hold it! Saturn's clouds are arranged in thick bands, though fainter than those found on Jupiter, and are mainly made up of ammonia and water ice. Another unique feature of this planet is its bright blue north pole, which is a swirling storm and hexagonal in shape. The blue appearance is caused by the way sunlight is scattered in the atmosphere.

The longest and most in-depth study of the planet Saturn was undertaken by the

Cassini spacecraft. From 2004 to 2017, the spacecraft not only observed Saturn but did numerous close fly-bys of some of its moons. This was no easy feat – Saturn has the most moons in our solar system, totalling 82! The spacecraft could only study a handful of moons, and at the time only 62 of Saturn's moons were known.

Some of the most important discoveries of Saturn's moon Enceladus were found by the *Cassini* spacecraft. Enceladus is similar to Jupiter's moon Europa, an ocean world with a surface of ice. Erupting from this surface are hundreds of geysers, which eject ice and gas from the subsurface oceans into space. The material from the geysers of this little moon – only one seventh the size of our Moon – actually forms one of Saturn's rings, the **E Ring**.

Incredibly, we can get a better understanding about Enceladus's oceans just by studying its geysers. *Cassini* found that the geysers contain chemicals that are considered the building blocks of life (large chains of atoms that include carbon, called organic molecules). Additionally, the spacecraft found evidence that the bottom of the ocean on Enceladus is in contact with its hot, rocky core. Similar environments can be found in the depths of Earth's oceans near hydrothermal vents, where bacteria and other microorganisms manage to survive without sunlight. Like the Earth, it is possible extremophiles may be able to survive in the depths of Enceladus's oceans.

Although Enceladus is a world of water and ice, Saturn's largest moon Titan couldn't be more different. Not only does Titan have a thick orange-coloured atmosphere, it is also the only object in our solar system (other than the Earth, of course) to have liquid lakes, rivers and seas on its surface. Surprisingly, these are not made from water, but are likely made from hydrocarbons such as methane and ethane, which are very similar to petroleum and crude oil.

As part of the *Cassini* mission to study Saturn's moons, a probe called *Huygens* travelled through Titan's thick atmosphere and landed on its surface; it was the first probe to ever land on a moon other than our own. The surface of Titan was described as having a texture similar to crème brûlée, meaning it had a frozen crust that could be broken with reasonable force. Even though the surface of Titan is incredibly cold, around -180°C, it may still be possible for life to exist there, either at an incredibly early stage on the surface or in a possible water ocean, which scientists suspect may exist underground.

Uranus

Although this planet might have an unfortunate sounding name, Uranus is actually the only planet named after an Ancient Greek deity instead of a Roman one. In Greek mythology, Uranus was the father of Saturn[13] (*Cronus* in Greek). Uranus spent many centuries hiding in plain sight and wasn't discovered until long after the previously discussed planets were. Even though it can be seen with the naked eye from dark locations, it was not recognised as a planet until it was observed through a telescope. Even John Flamsteed (1646–1719), the first Astronomer Royal at the Royal Observatory, Greenwich, sighted the planet six times as early as 1690, but mistakenly classified it as a star in the constellation Taurus, due to its distance from the Sun and slow orbit.

It wasn't until 1781, when observing the planet from his telescope in Bath, England, that Sir William Herschel (1738–1822)

---

[13] If the Roman tradition had been followed, Uranus would have been called Caelus. This would have probably led to fewer sniggers from audiences when astronomers attempt to talk about this planet seriously...

realised that Uranus wasn't a star. Herschel was stumped, he had mislabelled Uranus as a comet since he thought the 'star' was moving across the sky. After a two-year, concerted effort from astronomers around Europe to gather data, Herschel finally concluded that Uranus was a planet (third time's a charm!) However, this planet has never had good luck with names. Herschel originally named the planet 'George's Star' after King George III, the British monarch at the time. As you can imagine, this name wasn't popular outside of Britain, particularly with the American Revolutionary War happening alongside the discovery, so the name was changed to keep within the tradition of Classical mythology.

Despite its unusual history, this is an underrated planet that deserves some love and attention. Uranus is one of two ice giant planets in our solar system. Unlike gas giants, ice giant planets are mainly made up of 'ices'. This is a little misleading because ice giant planets are not covered in ice and have no definite surface. In astronomy, ices are not like sheets of ice that we experience on Earth, but instead they are volatile elements that have freezing points above -173°C. Below an ice giant's gaseous atmosphere is a thick mantle of ices like ammonia (NH_3), water (H_2O) and methane (CH_4). Ices are not necessarily cold and the mantles of the ice giants are sometimes referred to as water-ammonia oceans because the ices are more of a texture similar to a hot, dense fluid.

Uranus has a beautiful, pale-bluish hue that comes from methane gas in its atmosphere. The colouring of the planet suits its temperature, since the atmosphere of Uranus makes it the coldest planet in the Solar System, even though it isn't the furthest away from the Sun. Temperatures in its atmosphere have been known to dip down to a chilly -224°C.

In some sense, Uranus is a bit of a misfit in the Solar System because it is the only planet that is tilted sideways, meaning it appears to orbit the Sun like a ball rolling across the floor. It is possible that Uranus was struck by an early planet that caused it to tilt over on its side billions of years ago. Uranus's faint rings are also sideways, sitting vertical to the planet, looping over its north and south poles. Similarly, its 27 moons orbit the planet in the same way. These moons break the tradition of Greek and Roman mythology and are named after characters created by William Shakespeare and Alexander Pope.

Neptune

The last planet of our solar system, Neptune, is too faint to be seen in our night skies with the naked eye. Like Uranus, Neptune was first mistaken for a star in early charts, owing to its slow orbit, including by Galileo Galilei in his observations of 1612. Neptune wasn't discovered by a telescope though, but through mathematical predictions.

In 1846, around 65 years after the discovery of Uranus, the astronomer and mathematician Urbain Le Verrier (1811–1877) was able to predict the existence of Neptune by observing the movement of Uranus. As Uranus orbits the Sun, gravity pulls it towards Neptune when the two planets are near one another, which perturbs Uranus's orbit. Le Verrier was able to meticulously comb through observations of Uranus through its 84-year orbit and calculate the possible location of a planet beyond its orbit. The new planet was confirmed by telescope observations at the New Berlin Observatory.

Neptune was named after the Roman god of the sea, which fits the planet's striking, deep-blue colouring. Like Uranus, Neptune is also an ice giant, smaller than the gas giants Jupiter and Saturn but larger than the rocky planets. Both of the ice-giant planets are relatively similar in size, though Neptune is slightly smaller than Uranus in diameter. Methane in the atmosphere is what gives this planet its vivid colour, as methane gas absorbs the lower energy red light so that the planet appears blue.

Neptune features dynamic storms and its high winds, up to 2,100 kilometres per hour (1,300 miles per hour), make it the windiest planet in the Solar System. Similar to Jupiter's Great Red Spot, Neptune has large circular storms that tend to be deep blue in colour. These storms do not appear to last as long as the storms on Jupiter, though past Neptunian storms have been named by astronomers, including the Great Dark Spot and the Small Dark Spot[14].

---

[14] Even astronomers sometimes run out of creative names for the variety of phenomena in space. Self-explanatory or obvious names are the way to go when inspiration fails to strike!

This planet also hosts 14 moons, the largest and most substantial of which is Triton. Triton is an unusual moon as it orbits Neptune backwards. All of the planets in our solar system orbit the Sun in the same direction that the Sun spins on its axis. Similarly, most of the moons in the Solar System will orbit their respective planets in the same direction. Astronomers speculate that Triton may have started out as a planet or dwarf planet that was captured by Neptune's gravity. This may explain why it has a backward, or **retrograde**, orbit.

Being so far away from the Sun, the surface of Triton is covered in ice, including nitrogen, water and carbon-dioxide ice, and it has an atmosphere made up mainly of nitrogen. Volcanoes are scattered across the surface of the moon, known as cryovolcanoes, which erupt lavas that are made up of water and other ices (such as ammonia). This moon also has geyser-like eruptions of nitrogen gas and dust, which resembles other places with volcanism, like Earth and the moons Io, Europa and Enceladus.

Both the ice giants Uranus and Neptune still hold many mysteries. To date, only NASA's *Voyager 2* spacecraft has been able to study these two planets with fly-bys in the 1980s. The distance to these planets makes them difficult to reach, since Uranus is two times further away from the Sun than the planet Saturn, and Neptune is over three times further away!

Everything Left Over: Dwarf Planets, Comets, Asteroids and Meteoroids

While this book has explained *how* planets form and explored the features of the planets in our solar system, one thing that hasn't been discussed is the obvious question: what makes a planet a planet? It may seem odd to have left this question until one of the last sections, but it is best answered by taking a look at the dwarf planets and comparing them to the eight planets in our solar system to see why they are different.

It has been difficult for scientists to agree upon a definition for what constitutes a planet, because objects in space are not easily placed in neat categories. While it might be easy to tell the differences between two individual objects, it is difficult to determine the boundaries between more general categories.

It wasn't until an International Astronomical Union (IAU) meeting in 2006 that astronomers voted in favour of a definition of a planet, which led to Pluto being reclassified as a dwarf planet. This decision was controversial even in the astronomy community, but it was a decision with some reasoning behind it.

The definition was decided as follows. To be a planet, the object needs to fulfil three requirements:

1. It needs to orbit the Sun.
2. It needs to be spherical, like a ball.
3. It needs to have a defined and cleared orbit around the Sun.

Seems simple, right? For many objects, like Pluto, the first two requirements are easily met. All objects in our solar system are in orbit around the Sun. Similarly, for an object in space to be spherical, it needs to be massive enough so that it is pulled together and rounded by gravity, a situation known as **hydrostatic equilibrium**.

Pluto unfortunately fails on the third requirement of the definition because it is found in the middle of the Kuiper belt, a place similar to the asteroid belt but beyond the orbit of Neptune. The mass of Pluto is only a fraction of the total mass of icy and rocky objects in the path of its orbit around the Sun. Only objects with high mass can clear their orbits of debris with their stronger gravitational pull. Objects that meet the first and second criteria but not the third are now known as dwarf planets. Being out in the Kuiper belt means it takes Pluto 248 years to orbit the

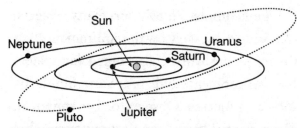

Figure 3: The orbits of the giant planets around the Sun seen in comparison to the orbit of Pluto. Pluto's orbit is inclined compared to the other eight planets in the Solar System.

Sun. Its orbit is also not in the same plane as the other planets, but on an incline (as seen in Figure 3).

But don't feel too bad for Pluto, there are other objects out in the Kuiper belt to keep this little dwarf planet company! Pluto has five moons (the largest is Charon) and there are other nearby dwarf planets including Eris, Haumea and Makemake[15].

---

[15] The Kuiper belt isn't the only place with dwarf planets. There is also a dwarf planet in the asteroid belt called Ceres. In total, there are only five official dwarf planets in our solar system.

Like Neptune, Pluto is too far away from the Sun to be seen by the naked eye. This dwarf planet wasn't discovered until 1930 by astronomer Clyde Tombaugh (1906–1997) at Lowell Observatory in America. Pluto was detected by astrophotography, where Tombaugh compared photographs of the night sky over time and was able to see the slow movement of the dwarf planet across the sky. Since it was originally thought to be the ninth planet in the Solar System, Pluto was named after the Roman god of the underworld. The name was fitting for the dwarf planet that evaded detection for so long because this Roman god was also able to make himself invisible using a magical cap. Pluto was actually named by Venetia Burney, an 11-year-old from Oxford, England, after the astronomers at the Lowell Observatory unanimously voted on her suggestion when attempting to name the newly discovered object.

When NASA's *New Horizons* spacecraft first reached Pluto in 2015, it became the only spacecraft so far to study the dwarf planet up close. *New Horizons* found that there was surprisingly more to this little world than what anyone originally thought. Like the rocky planets, Pluto does have a surface, but it is mainly made up of nitrogen ice with temperatures that dip to -233°C. The colour of the surface varies from a deep black to orange and white. Geographical features on this planet include plains, cryovolcanoes (like Saturn's moon Enceladus), mountains made of water ice, and glaciers. The black regions are very likely made from tar from hydrocarbons, the basis of crude oil and coal. The atmosphere surrounding Pluto resembles its surface. It is made up of nitrogen, methane (NH_3) and carbon monoxide (CO) gases.

When *New Horizons* sent back images of Pluto, they showed a frozen world with plains and mountains arranged to resemble what looks like a giant heart on the surface of the dwarf planet. It was almost like little Pluto, which is smaller than our own Moon, was sending a message out to us, thanking us for not forgetting its place in our solar system (I'm not crying, you're crying).

There are a variety of other things in our solar system that are not dwarf planets, planets or moons. Asteroids, comets and meteoroids are the building blocks of planets. These objects are leftovers from the dusty disc that once orbited our Sun when planets first formed.

Asteroids are large, rocky bodies made up of oxygen, silicon and metals, like iron. They can be anywhere from a few metres in size up to 1,000 km (just under the length of Great Britain). Asteroids are literally pieces of planets that never had the opportunity to assemble together. Smaller rocky bodies in space are known as meteoroids, which are anywhere from the size of small dust grains up to a couple of metres. Most of these asteroids and meteoroids in our solar system can be found in the asteroid belt between Mars and Jupiter. Sometimes these asteroids can collide with one another and knock each other off course and out of the asteroid belt altogether. The strong gravitational pull from the mammoth-sized Jupiter disrupts these rocky bodies in the asteroid belt and prevents planets from forming there.

In many novels and films about space travel, passing through an asteroid belt is highly dramatic and filled with suspense. Spaceships weave in and out of asteroids, often narrowly escaping an oncoming attack. Sorry to be *that* person, but this is a good example of bad astronomy! In asteroid belts, including our own asteroid belt, the asteroids are extremely far away from one another, relative to the rest of our solar system. Asteroids are separated from one another by hundreds of thousands of kilometres. Space is a *big* place, much bigger than we can comprehend in the context of our everyday experiences. This means that if you do ever commandeer an alien spacecraft and travel through the Solar System, there is no need to dodge oncoming asteroids as you pass through the asteroid belt (but, it does make good cinema).

Comets are the large, dirty snowballs of space. These frozen objects are made mainly out of dust, ice and small rocks. Comets, like everything in our solar system, orbit the Sun. The shape of their orbits are highly elliptical, meaning they spend most of their time far away from the Sun. In the event that a comet passes close to the Sun, its ice will begin to evaporate and a fuzzy cocoon of gas will surround the comet, called a coma[16]. On the

---

[16] 'Coma' comes from Ancient Greek and translates to 'hair'. The material evaporating and streaming off the comet in sunlight is thought to be like hair blowing in the wind.

opposite side to the sunlit side, the comet will extend a bright gas tail and a fainter tail made up of dust.

Comets are scattered throughout the Solar System. Like Pluto and most of the other dwarf planets, they originate in the Kuiper belt. These comets orbit the Sun in under 200 years, so they are known as short-period comets. Comets can also come from the very edges of the Solar System, in a mysterious region called the Oort Cloud. While the Oort Cloud may sound like something straight out of science fiction, this is a region filled with icy space debris that surrounds the Sun a little like a spherical shell. This debris is thousands of times further away from the Sun than the Earth, meaning the Sun is hundreds of millions to billions of times fainter to these objects. Comets originating from this region are known as long-period comets because it can take them thousands of years to orbit around the Sun.

Asteroids and comets create beautiful phenomena in our solar system, called meteor showers. When a piece of an asteroid, a meteoroid or a comet falls towards the Earth, it creates spectacular streaks across the sky as it burns up in the atmosphere, which is more commonly known as shooting stars or meteors. While it may seem like our solar system is stagnant and stationary, from here on Earth, these marvellous meteor showers occurring throughout the year are a reminder that everything around us is moving and changing, even if we cannot feel it.

All of these leftovers – the icy and rocky debris – gives us a better understanding of what our early solar system was like because these objects have remained relatively pristine since its formation. Their environments aren't altered by geological activity over the past billions of years. Astronomers have started to do what previously seemed like the impossible, studying these environments by sending spacecraft to orbit and land on this moving space debris. Space missions like *Hayabusa*, *Hayabusa 2* and *OSIRIS-Rex*, which travelled to asteroids, and the *Rosetta* mission, which landed on a comet, have allowed these environments to be studied up close for the first time.

Alien Worlds: Our Search for Exoplanets

In the 1990s, astronomers discovered that our solar system was not alone in the Galaxy. Out of the hundreds of billions of stars living alongside the Sun, there were stars that have their own planets in orbit around them, called extrasolar planets, or exoplanets for short. The discovery of the first exoplanet orbiting a main-sequence star (a star that is in the prime of its life) in 1995 by astronomers Michel Mayor and Didier Queloz was awarded half of the 2019 Nobel Prize in Physics.

This famous exoplanet is in orbit around 51 Pegasi, a Sun-like star in the flying-horse constellation called Pegasus. The exoplanet, 51 Pegasi b, or 'Dimidium', is around 50 light years away from our solar system. Dimidium is a hot Jupiter exoplanet, or a Jupiter-sized gas planet that is incredibly close to its star and therefore is extremely hot. While this exoplanet is only around half the mass of Jupiter, it is nearly ten times closer to 51 Pegasi than Mercury is to our Sun and only takes four Earth-days to complete one orbit around its star! The temperatures of this gas giant at its close distance may be as high as 1,000°C. While water was found in the atmosphere of 51 Pegasi b, it is unlikely that life exists because of its extreme temperature.

The discovery of 51 Pegasi b marked the first exoplanet in orbit around a Sun-like star. However, the first confirmed exoplanet was actually observed three years prior, in 1992. The exoplanet was found in orbit around a pulsar PSR 1257+12, more recently named Lich by the International Astronomical Union. Sometimes known as neutron stars, pulsars are often thought to be dead stars. They are the remnants of massive stars that have slowly run out of fuel over time and erupted as supernovae. The pulsar that remains after this explosion is the leftover core of the star, which has a fast spin, high density and strong magnetic field. These stars are so dense that a teaspoon size chunk from a pulsar on the Earth would weigh as much as Mount Everest.

The two exoplanets initially found in orbit around the pulsar Lich were named Poltergeist and Phobetor. These exoplanets are about four times the mass of the Earth and have orbits similar in size to Mercury's orbit around the Sun.

Another exoplanet was found orbiting the pulsar in 1994, named Draugr[17], which is half the mass of Mercury and twice as close to the pulsar. The intense radiation from pulsars mean that it would be unlikely for life to exist on an exoplanet in orbit around one. However, this first discovery was an important step in understanding how common exoplanets are in the Universe.

There are a variety of ways to detect exoplanets, though many of them do not

---

[17] Most stars and planets are given impossibly long names, filled with letters and numbers. This planetary system is an excellent example of why astronomers need to be more creative when they name new astronomical objects. The pulsar Lich is named after a fictional undead creature that can control things with magic. A poltergeist is a noisy ghost in folklore. Phobetor is a supernatural beast that appeared in dreams, found in Roman poet Ovid's writings. Lastly, Draugr is an undead creature from Norse mythology. Perfect names for a planetary system orbiting a dead star!

involve photographing or directly imaging the planet in its orbit around the star. Planets do not emit light like the stars they orbit, but they can reflect starlight. This is the same reason we can see the planets in our own solar system, light from our Sun is reflecting off them. Compared to the light of the star, the light reflecting off the exoplanet is very small. Directly imaging a faint exoplanet around its star is like finding the light from a single candle in a blazing inferno. It isn't impossible and has been done in the past, but it is very difficult.

Instead of imaging, indirect methods must be used to find these exoplanets hiding in the light of their stars. 51 Pegasi b was found using the radial-velocity method, more commonly known as the 'wobble method'. If a planet is massive enough, a planet similar in size to Jupiter, for example, the gravity from both the star and the planet will pull on one another. While the planet orbits the star, the planet's gravity will tug on the star and cause the star to wobble in a circle.

Extraordinarily, this rocking motion can be detected by telescopes! As the star moves back and forth, the light from the star shifts in colour. As the star moves away from the Earth, its light becomes redder. As it moves towards the Earth, the light becomes bluer. This is called the Doppler Effect[18] or 'redshift' and 'blueshift'. These changes in the star's light can be observed by telescopes. From these observations, we can even calculate the distance of the planet to the star and the mass of the planet using our understanding of planetary motion and gravity that Johannes Kepler and Sir Isaac

---

[18] The Doppler Effect not only changes the colour of light, but also the pitch of sound. Police sirens work in the same way. As the sound travels towards you, the pitch is higher. When the sound travels away from you, the pitch is lower.

Newton originally proposed at the turn of the 18th century.

The wobble method of finding exoplanets can detect massive planets that are close to their stars because the gravitational interaction between the planet and star is greater. For detecting other exoplanets, including less massive ones, the transit method can be used (see Figure 4). Like Mercury and Venus, which from Earth appear to pass occasionally in front of the Sun, or transit the Sun, exoplanets can transit the stars they orbit. When an exoplanet passes in front of its star, it will block some of the star's light and cause it to dim. This dip in brightness can be observed, which repeats every time the planet moves in front of the star. From the transit, the distance of the planet from its star and its size can be determined.

The majority of exoplanets have been found using the transit method. Missions

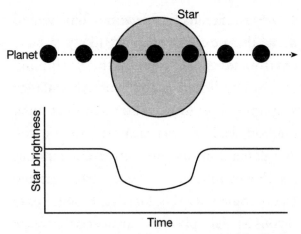

Figure 4: An example of the transit method for finding exoplanets. The exoplanet will pass in front of its star and block out the star's light. This decreases the star's brightness for a short amount of time before returning to its original brightness.

like NASA's *Kepler* space telescope and the detection programme Wide Angle Search for Planets (WASP) were instrumental in finding new exoplanets. While WASP was primarily able to detect larger planets (Neptune- and Jupiter-sized), *Kepler* was designed to search for Earth-like exoplanets that are not only a similar size to our planet, but also in their solar system's habitable zone. This would make these exoplanets more likely to have environments like the Earth, potentially possessing liquid water and capable of sustaining life. An advantage of the transit method is that the atmosphere of the exoplanet can also be better understood. As the exoplanet travels in front of the star, some of this starlight will pass through the planet's atmosphere. The different gases making up the atmosphere will absorb some of this light. By studying the light that has been absorbed by the atmosphere, astronomers can actually determine what the exoplanet's atmosphere is made up of, which gives us a better idea of the nature of these alien worlds.

While the transit method allows us to study the environment of the exoplanet, it is highly dependent on the orientation of the Solar System that is being observed. If the exoplanet doesn't pass in front of their stars from our vantage point here on the Earth, then the exoplanet won't be detected using this method.

Other methods for detecting exoplanets include gravitational microlensing (see Figure 5). Gravitational microlensing is a consequence of Albert Einstein's (1879–1955) theory of general relativity. At large scales, there are some bizarre effects in physics, where space doesn't act in a way that fits our experience on the Earth. In the cosmos, general relativity tells us that anything with mass – a galaxy, a star, a planet, even a human being – warps the very fabric of space and time. If there is an object, like a star or galaxy, that is far away from the Earth, then the light from that distant object can be bent by another object which happens to be between the Earth and the more distant object. This is known as the gravitational lens effect, because the object in between the Earth

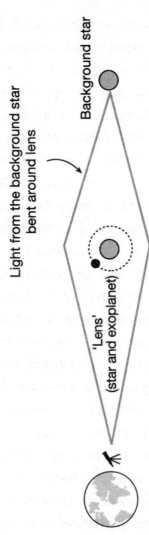

Figure 5: How microlensing works. The star and exoplanet act like a lens and bends the light from a background star around it. From Earth, telescopes can pick up the change in the background star.

and distant object acts a little like an optical lens. It might seem strange, but this phenomenon is similar to how the bottom of a wine glass warps a pattern underneath it. Instead of a wine glass, there is a massive body in space that acts like the lens. As a background star passes behind the star and exoplanet, the light from the background star will brighten temporarily. We can calculate the mass of the exoplanet in orbit around the star that acts like the lens from the brightening of this background star.

Gravitational microlensing is incredibly useful for detecting planets that have lower masses, down to a couple of Earth-masses. It can also be useful for finding exoplanets that are far away from the Earth, tens of thousands of lightyears away from our solar system! The problem with this technique is that the planets found from microlensing can only be observed once. It is rare that a background star would pass directly behind the same exoplanetary system again. This makes it difficult to confirm these serendipitous observations of exoplanets that are detected through microlensing.

Lastly, astrometry can be used to search for exoplanets. Remember how an exoplanet can tug on the star it is orbiting and cause it to wobble about in space? Astrometry is when the small movements of the wobbling star are compared to the positions of other stars to determine that an exoplanet is causing a star to change positions over time. If you thought microlensing was hard to do, try astrometry. The wobbles are incredibly small and nearly impossible to detect from the Earth because of our atmosphere, which bends and distorts light coming from space. However, the *Gaia* spacecraft in orbit around the Earth is observing more than a billion stars with the goal of detecting new exoplanet systems using this method.

So far, astronomers have detected some intriguing worlds since the first exoplanets were found. Kepler-16b is a Saturn-mass exoplanet that is orbiting not one, but two stars. If you could stand on it, then the sky might look something similar to the planet Tatooine in *Star Wars* with its two suns. Kepler-10b is a few times more massive than the Earth, but over 20 times closer to its star than Mercury is to our Sun. While this is a rocky exoplanet, its surface is likely scorched by its star and its day side will get to temperatures hot enough to melt iron. One of the most exciting and largest planetary systems that has been found is around the TRAPPIST-1 star, which has seven rocky planets. Out of these planets, five are similar to the size of the Earth and three are within the star's habitable zone. TRAPPIST-1 is a red dwarf star, which is the coolest and smallest type of star. To be in TRAPPIST-1's habitable zone, the exoplanets need to be much closer to the star than in our own solar system, in order to be warm enough for liquid water to exist on their surfaces.

Currently, over 4,000 exoplanets have been detected, the majority of these detected by *Kepler*. This telescope detected these exoplanets in a portion of the sky around the Summer Triangle star pattern, visible in the northern hemisphere during the summer months. This pattern consists of three bright stars in the shape of an upside-down triangle and is easy to spot even from cities. The exoplanets were found between the stars at the top of the triangle, Deneb and Vega. Next time you see these stars in the sky, you may want to pause and wave. After all, there could be someone out there watching us, just like we are searching for them.

A Final Note: Is There Life in the Universe Other Than on Earth?

It is natural to look up at the stars and wonder if we are alone. From our studies of young stars, we know that it is possible that there will be on average one exoplanet per star. With on order of 100 billion stars in the Milky Way galaxy, this means that there could be 100 billion exoplanets in our galaxy as well. We've just not detected all of these planets at this point in time. There will be even more exoplanets in our universe because there are more galaxies out there. Current estimates show that there are on the order of 1 trillion galaxies out in the cosmos, or 1,000,000,000,000 galaxies. If each of those galaxies has about the same number of stars (and each of those stars has on average one planet in orbit around them), then that would mean that there are around 10^{23} exoplanets in our universe, or about 100,000,000,000,000,000,000,000,000 exoplanets. In all probability, it appears that life beyond the Earth exists. The problem is finding what could be a needle in a haystack when the haystack is the size of the Grand Canyon. The Universe is a massive place!

While we would love to search for life on exoplanets, a serious difficulty we face is how far away these exoplanets are. Our closest exoplanet is called Proxima b, it orbits the star Proxima Centauri in the southern constellation Centaurus. This exoplanet is a little over four lightyears from the Earth. With our modern spacecraft, for example the *Parker Solar Probe* which reaches 690,000 kilometres per hour at its top speed, it would take us over 6,600 years to reach this exoplanet!

Instead of focussing on exoplanets, we can continue exploring our own solar system because we are able to access it and study it up close with spacecraft much more easily. If we want to search for life, the best places to start are the locations we find liquid water because every form of life on the Earth needs liquid water to survive. For example, the icy moons around Jupiter and Saturn – Europa, Ganymede and Enceladus – all have evidence of liquid water oceans below their icy surfaces. In addition to moons, there is evidence of underground lakes near the poles of the planet Mars and an internal ocean below surface ice on the dwarf planet Ceres in the asteroid belt. Even without these large bodies of water present, it is possible that life which is nothing like the kind we find on the Earth exists in less obvious places in the Solar System. Saturn's moon Titan does not have liquid water on its surface due to its cold temperatures, but it does have lakes and rivers on its surface that are made of liquid methane where novel forms of life may reside. Even the planet Venus has been theorised to have extreme forms of life in its milder upper atmosphere.

The important thing we must remember is to keep searching. Life is out there, and it's likely just a matter of time before we find it. From our solar system to beyond, astronomers have their work cut out for them to discover new planets and explore the possibilities for life on these alien worlds.

Glossary

Absolute zero – the lowest possible temperature in the Universe. On the Kelvin temperature scale, it is listed as 0 kelvin. In Celsius, it is -273.15°C.

Asteroid – a rocky object in space. Asteroids are mainly found in the asteroid belt and the Kuiper belt.

Asteroid belt – a ring of asteroids in our solar system between the orbits of Mars and Jupiter. It is possible that other stars have their own asteroid belts.

Astrometry – a way to search for exoplanets by searching for the small movements of a star that is being influenced by an orbiting planet against other nearby stars. More generally, astrometry is the precise measurements of the positions and movements of stars and other astronomical objects.

Blueshift – caused by the Doppler effect. When an object in space, like a star, moves towards the Earth, its light will become bluer.

Comet – an object in space that is mainly made up of ice and dust. Comets can be classified as 'short-period' and orbit the Sun in under 200 years, or 'long-period' and orbit the Sun in thousands of years.

Doppler effect – as objects move in space, their light will become redder or bluer depending if they are travelling away from or towards the Earth.

Dwarf planet – a body in space that acts like a planet but does not meet the criteria to be classified as one. Dwarf planets in our solar system are Pluto, Eris, Haumea, Makemake and Ceres.

Exoplanet – also called an extrasolar planet. Exoplanets are planets that orbit other stars in space and are a part of their own solar systems.

Extremophile – forms of life (like microbes or microscopic organisms) that are able to survive and thrive in extreme environments, including extremely hot and cold temperatures.

Frost line – also called the 'snow line'. This is an imaginary line that represents the distance from a star where it is cool enough for gases or liquids to begin freezing into ice grains.

Gas giant planet – a giant planet made mainly of hydrogen and helium gases. In our solar system, Jupiter and Saturn are gas giant planets.

General relativity – a well-studied scientific theory published by Albert Einstein in 1915. This theory describes gravity as the curvature or distortion of space and time.

Habitable zone – sometimes called the 'Goldilocks Zone'. This is the region around a star where the temperature is mild enough for liquid water to exist on the surface of an Earth-like planet.

Heliocentric – the accepted model of our solar system where the Sun is in the centre and the planets orbit the Sun. The alternative is 'geocentric', or the past model that the Earth is at the centre of our Solar System.

Ice giant planet – a giant planet made mainly of **ices** that are heavier than hydrogen and helium.

Ices – a relatively misleading term in astronomy because ices are not necessarily cold. Ices are made up of volatile elements with freezing points above -173°C and include ammonia, water and methane. Ices can be incredibly hot and fluid-like in nature.

Kinetic energy – the energy objects will have due to their motion.

Kuiper belt – this is a part of the Solar System that can be found beyond the orbit of Neptune. It is filled with comets, asteroids and other bodies (including four of the five dwarf plants).

Lightyear – the distance light takes to travel in a year, or 9.5 trillion kilometres (nearly 6 trillion miles).

Main-sequence star – stars where hydrogen is being fused into helium is occurring. Most stars in the Universe are main-sequence stars (including our Sun).

Meteor – sometimes called 'shooting stars'. Meteors are normally pieces of comets, asteroids or meteoroids that burn up in Earth's atmosphere and appear as a streak of light across the sky.

Meteorite – a piece of space debris (i.e. a meteoroid, asteroid or potentially a comet) that falls onto the surface of the Earth. Chondrite meteorites are made up of chondrules, or small, round dust particles.

Meteoroid – rocky bodies that are smaller than asteroids in space.

Microlensing – a way to detect exoplanets. Gravitational microlensing relies on Albert Einstein's theory of general relativity. A star with an exoplanet will act like a 'lens' that bends the light from a passing background star. The light from this background star will brighten temporarily and can be used to find this exoplanet orbiting the star that acts like a lens.

Moon – a natural satellite of a planet, meaning moons orbit planets. While planets and dwarf planets orbit the Sun, moons orbit planets.

Nebula – a region in space made up of dust and gas where stars are forming.

Nuclear fusion – the process where elements (like hydrogen) are fused together to form heavier elements (like helium). Hydrogen fusing into helium is what is powering all main-sequence stars.

Oort Cloud – the outer region of our solar system that is filled with icy space debris and surrounds the Solar System like a spherical shell. Long-period comets are thought to come from the Oort Cloud.

Protostar – the precursor to a star, or a 'baby' star.

Pulsar – also known as 'neutron stars'. Pulsars are the remnants of massive stars that ran out of fuel and erupted as supernovae. The leftover core of the star collapses and becomes a neutron star. Pulsars have a fast spin, high density and strong magnetic field.

Radial-velocity method – a way to detect exoplanets in orbit around other stars, sometimes called the 'wobble method'. An exoplanet is detected by observing the change in colour of a star's light that is caused by the movement of the star as the exoplanet orbits it.

Red dwarf star – the coolest and smallest type of star.

Redshift – caused by the Doppler effect. When an object in space, like a star, moves away from the Earth, its light will become redder.

Terrestrial planet – a terrestrial planet is the same as a rocky planet, which include Mercury, Venus, Earth and Mars in our own solar system.

Transit method – a way to detect exoplanets in orbit around other stars. As an exoplanet passes in front of their stars, they will block out a small amount of star light. This dip in the brightness of the star can determine an exoplanet is present.

Royal Observatory Greenwich Illuminates

Stars
by Dr Greg Brown
978-1-906367-81-7

Space Exploration
by Dhara Patel
978-1-906367-76-3

Black Holes
by Dr Ed Bloomer
978-1-906367-85-5

The Sun
by Brendan Owens
978-1-906367-86-2